**Editors**
Lorin Klistoff, M.A.
Kathleen "Casey" Petersen

**Managing Editor**
Karen Goldfluss, M.S. Ed.

**Illustrators**
Mark Mason
Renée Christine Yates

**Cover Artist**
Brenda DiAntonis

**Art Manager**
Kevin Barnes

**Art Director**
CJae Froshay

**Imaging**
Alfred Lau
Rosa C. See

**Publisher**
Mary D. Smith, M.S. Ed.

# Author

*Bev Dunbar*

(Revised and rewritten by Teacher Created Resources, Inc.)

This edition published by **Teacher Created Resources, Inc.**
6421 Industry Way
Westminster, CA 92683
www.teachercreated.com

**ISBN-1-4206-3535-2**

©2005 Teacher Created Resources, Inc.

Made in U.S.A.

The classroom teacher may reproduce copies of materials in this book for classroom use only. The reproduction of any part for an entire school or school system is strictly prohibited. No part of this publication may be transmitted, stored, or recorded in any form without written permission from the publisher.

# Table of Contents

**Introduction** .................................................. 3
**How to Use This Book** .................................. 4
**Exploring Coins** ........................................... 5
    What Is Money? ......................................... 6
    What Is a Shop? ......................................... 7
    Let's Make a Class Shop. ............................ 9
    What Is a Coin? ........................................ 11
    Trace It, Rub It, Press It. ........................... 14
    Match It .................................................. 15
    Sort Me ................................................... 18
    Coin Bingo. .............................................. 20
    Spin a Coin ............................................. 22
    Hide a Coin ............................................. 24
    Coin Concentration .................................. 25
    Which Is Worth More? .............................. 27
    Checkup ................................................. 28
**Using Coins.** ................................................ 29
    Count with Me ......................................... 30
    Piggy Banks ............................................ 31
    Match My Card ....................................... 33
    Stack Them Up ....................................... 39
    Mix Them Up. .......................................... 42
    Supermarket Challenges. ......................... 43
    Find My Match. ....................................... 46
    Class Shop ............................................. 52
    Shopping with Up to 50¢. ......................... 54
    Shopping with Up to $1.00 ....................... 57
    Cash Registers ....................................... 60
    Who Spent More? .................................... 63
    Find the Same Amount ............................ 65
    What's My Change? ................................. 68
    Change Challenges. ................................ 72
**Exploring and Using Bills** ........................... 77
    What Is a Bill? ......................................... 78
    Make Your Own Bills ............................... 79
    Which Is Worth More? .............................. 83
    Make Your Own Catalog .......................... 86
    How Much Is That? .................................. 87
    Pocket Money ......................................... 89
    Lots of Money ......................................... 93
**Skills Record Sheet** ................................... 94
**Sample Weekly Lesson Plan** ..................... 95
**Blank Weekly Lesson Plan** ........................ 96

# Introduction

Money is one mathematical topic that is familiar to all students. You discuss money regularly as part of your class routines and news or when calculating the total collected for a charity, a special excursion, or lunch costs at the school cafeteria.

*Math in Action: Money* is a companion to the other number books in the Math in Action series. Included in this book are many action-packed ideas for developing skills in recognizing, naming, matching, and using coins and bills in fun, practical ways. The activities can range from simple to super-challenging, to help you support different ability groups.

Making your teaching life easier is a major aim of this series. The book is divided into sequenced units. The units contain activity cards and worksheets for small groups or a whole class to explore. You also will find easy-to-follow instructions, with assessment help in the form of clearly stated skills linked to a record sheet (page 94).

Each activity is designed to maximize the way in which your students construct their own understanding about money. The activities are generally open-minded and encourage each student to think and work mathematically. The emphasis is always on mental recall, as well as the practical manipulation of coins and notes.

Have fun exploring money concepts with your students.

# How to Use This Book

## ❏ Teaching Ideas

Included in this book are many exciting teaching ideas which have been placed into three sections to assist your lesson planning for the whole class or small groups. Each activity has clear learning outcomes and easy-to-follow instructions. Activities are open-ended and encourage your students to think for themselves.

## ❏ Reproducible Pages

In this book are many reproducible pages. Below are some examples of the different types of pages which are included in this book.

### *Reusable Worksheets*

(e.g., page 8, What Is a Shop?)
These support free exploration, as well as structured activities. They are great for reuse with small groups.

### *Playing Cards*

(e.g., page 40, Stack Them Up)
Cut these out, shuffle, and use over and over again for small group games. Copy each set in different colors for easy classroom management.

### *Activity Cards*

(e.g., page 55, Shopping with Up to 50¢)
Use these as an additional stimulus in group work. The language is simple and easy-to-follow. Encourage your students to invent their own activity cards too. You can laminate them so that they last for years.

## ❏ Skills Record Sheet

The complete list of skills is available on page 94. Use this sheet to record individual student progress.

## ❏ Sample Weekly Lesson Plan

On page 95, you will find an example of how to organize a selection of activities from "Using Coins" as a five-day unit for your class. A blank weekly lesson plan is included on page 96 for your use.

# Exploring Coins

**In this unit, your students will do the following:**

- ❏ Recognize the role of money in our daily lives
- ❏ Recognize, name, and match 1¢, 5¢, 10¢, 25¢, 50¢, and $1 coins
- ❏ Describe and sort coins by design, color, shape, size, and value
- ❏ Order coins by value

(The skills in this section are listed on the Skills Record Sheet on page 94.)

# What Is Money?

**Exploring Coins**

### Skill
- Recognize the role of money in our daily lives

### Grouping
- whole class

### Materials
- piggy bank, wallet, purse with mixed coins and bills inside

### Directions
- Discuss with students what you might find inside a purse or wallet. (e.g., a comb, tissue, bus ticket, money)
- Ask students, "What is money?" (e.g., coins and bills) "Why do we have money? How do you use it?" (e.g., to buy something to eat, to pay for a visit to the doctor, to save for a special toy)
- Ask students, "Where might you find it?" (e.g., in a purse, wallet, handbag, money box, cash register at a shop, bank, ATM)

### Variations
- Discuss the money students may have with them at school.
- Ask students, "Why do you need money at school? What can you buy with it?"
- Discuss pocket money. Ask, "Who gets this? On what do you spend it? Where do you keep it?"
- Tell students to imagine that there is no money in our world. Discuss ways in which their lives would be different. How would they get things?

Exploring Coins

# What Is a Shop?

## Skill
- ❑ Recognize the role of money in our daily lives

## Grouping
- ❑ whole class

## Materials
- ❑ large color posters of different kinds of shops and stores (optional)
- ❑ Shops worksheet (page 8)

## Directions
- ❑ Ask students, "What is a shop? Why do we have them? Why do you think shops have different names?" (e.g., grocer, butcher, pharmacy, café)
- ❑ Ask students, "Who works in each shop? What are special names for shopkeepers?" (e.g., the grocer)
- ❑ Ask students, "What sorts of shops do you go to with your family? Which is your favorite shop to visit? Why? What types of things can you buy there? Which shop don't you like to visit? Why? What types of things can you buy there?"

## Variations
- ❑ Have students make a class list of all the shops that were discussed. Ask them, "How many different shops altogether?"
- ❑ Have students make a class list of special shop vocabulary. (e.g., customer, shopkeeper, grocer)
- ❑ Use page 8. Have students color each picture. Then have them cut and paste the pictures onto a blank piece of paper. Beside each picture, have them draw items they could buy in each shop or have them cut and paste related pictures from magazines. Ask students, "How much money do you think these items cost?"

What Is a Shop? Exploring Coins

# Let's Make a Class Shop

## Skill
- ❑ Recognize the role of money in our daily lives

## Grouping
- ❑ whole class

## Materials
- ❑ tables, chairs, and shelves to make a class shop
- ❑ junk resources to sell in the shop

## Directions
- ❑ Ask students the following questions:
  
  "What sort of shop will we make in our classroom?"
  
  "What will it look like?"
  
  "What equipment do we need?""
  
  "What items do the shopkeepers need?" (e.g., a counter top, a cash register, products to sell)
  
  "What items do the customers need?" (e.g., a shopping basket, money, a shopping list)

- ❑ Have students collect all the suggested resources to make their class shop.

## Variations
- ❑ Have students make more than one class shop.
- ❑ Use page 10. Discuss what sort of shop the children might like to own. Ask, "What items would go on each shelf?" Have them draw and color in items to sell on the shelves. Have students write or copy the name of this shop on the sign at the front of the counter.

Let's Make a Class Shop — Exploring Coins

*Exploring Coins*

# What Is a Coin?

## Skills
- Recognize, name, and match 1¢, 5¢, 10¢, 25¢, 50¢, and $1 coins
- Describe and sort coins by design, color, shape, size, and value

## Grouping
- whole class

## Materials
- plenty of 1¢, 5¢, 10¢, 25¢, 50¢, and $1 coins
- paper coins (page 12) for each student
- coin designs (page 13) for each student

## Directions
- Discuss why people have coins.
- Describe and discuss the designs students see on the front of each coin. (e.g., George Washington, Abraham Lincoln, Thomas Jefferson) Also, discuss the backs of the coins, especially the states on the back of quarters.
- Describe and discuss the size of each coin.

   (e.g., Which coin is the largest? The smallest? Why do you think they are different sizes?)

   (e.g., Why do you think that the dime, which is worth 10¢, is smaller than the nickel, which is worth 5¢, or the penny, which is worth 1¢?) *(The dime actually used to be made of silver and the nickel made out of nickel. The dime was made smaller because its metal was more valuable.)*

- Discuss the color and value of each coin.

   (e.g., Why do you think they are different colors? Which coin is a different color than the others? Of what metal is that coin made? What number can you see? Why do they have different numbers on them?)

## Variations
- Have students cut out some mixed paper coins. Have them color the paper coin and match them with real coins. Have students glue their coins onto paper.
- Have students look at the coin designs. Have them cut out, color, and match to a real or paper coin.
- Have students discuss what each coin looks like on the back.

**What Is a Coin?**                                                                                  **Exploring Coins**

**What Is a Coin?**     **Exploring Coins**

**Exploring Coins**

# Trace It, Rub It, Press It

## Skills

- Recognize, name, and match 1¢, 5¢, 10¢, 25¢, 50¢, and $1 coins
- Describe and sort coins by design, color, shape, size, and value

## Grouping

- small groups

## Materials

- plenty of 1¢, 5¢, 10¢, 25¢, 50¢, and $1 coins or play money
- paper, pencils, crayons
- play clay

## Directions

- Have students investigate the shape and size of each coin by tracing around the edge of a real coin. Ask students, "Can you recognize each coin just by looking at the traced outline?" Have them copy the design on the inside of their tracing. Have them exchange their drawing with a friend. Ask, "Can you both recognize the coins you see?"
- Have students place coins under paper and rub the surface with a crayon. Discuss key features of each design which appear in the rubbings. Ask, "Why don't all coins have the same design?"
- Have students press coins into play clay. Have them each ask a friend to guess their coins just by looking at the impression.

## Variations

- Have students make a coin pattern. Have them each ask a friend to continue their patterns.
- Have each student trace the outline of a coin and create their own designs inside.
- Have students copy their coins by drawing them freehand. Have them make them as large or as small as they like. Have them see if their friends recognize each coin just by looking at their drawings.

#3535 Math in Action

14

©Teacher Created Resources, Inc.

**Exploring Coins**

# Match It

## Skills

- Recognize, name, and match 1¢, 5¢, 10¢, 25¢, 50¢, and $1 coins
- Describe and sort coins by design, color, shape, size, and value

## Grouping

- small groups

## Materials

- small items from the class shop
- Match It workstrips (pages 16 and 17—each cut into five strips)
- paper, pencils, crayons

## Directions

- Have students look at different items from the class shop. Ask them, "What prices would you like them to be?"
- Have students select prices from 1¢, 5¢, 10¢, 25¢, 50¢, and $1.
- Have students look at the item for sale on the strip.
- Have students decide how much it will cost (select a value from one of the coins).
- Have students copy the number onto the price tag.
- Have students exchange strips with a partner.
- Have students circle or color a coin that matches the price tag.

## Variations

- Have students match the prices on their Match It strips using real coins.
- Have students make up their own Match It strips by tracing or drawing different coins and items for sale.
- Ask students, "Can you find a different combination of coins to match each price tag?"

**Match It**             *Exploring Coins*

# Sort Me

## Skills

- Recognize, name, and match 1¢, 5¢, 10¢, 25¢, 50¢, and $1 coins
- Describe and sort coins by design, color, shape, size, and value

## Grouping

- small groups

## Materials

- a variety of coins or paper coins (page 12)
- Sort Me cards (page 19)

## Directions

- Have each student take a handful of coins. Ask them, "How many different ways can you sort these?"

    e.g., How many silver coins? Copper coins?

    How many five-cent pieces? Ten-cent pieces?

    How many large or small coins?

    How many coins with _____?

    How many coins with pictures of _____?

## Variations

- Have each student take a handful of coins. Have him or her turn over a Sort Me card. Have each student race to find and count all the matching coins in his or her pile. Ask students, "Who collected the most? How many coins like this altogether in your group?"

- Have students sort items from the class shop into groups according to price.

    (e.g., All the things that cost 10¢ go on this shelf.)

- Have students collect magazine cutouts, used wrappers, labels, or drawings of items in real life that cost exactly 1¢, 5¢, 10¢, 25¢, 50¢, and $1. Put these together into a large class book about money.

Sort Me Exploring Coins

| copper coins | silver coins |
| large coins | small coins |

Exploring Coins

# Coin Bingo

### Skills

- Recognize, name, and match 1¢, 5¢, 10¢, 25¢, 50¢, and $1 coins
- Describe and sort coins by design, color, shape, size, and value

### Grouping

- small groups

### Materials

- a variety of coins or paper coins (page 12)
- Coin Bingo boards (page 21)
- six coins (1¢, 5¢, 10¢, 25¢, 50¢, and $1) in a paper bag for the leader

### Directions

- Have each student take a Coin Bingo board.
- Have each student take any six coins and place them face up at random on his or her board. Students may have more than one of a certain coin.
- The leader takes a coin from the bag and calls out its value. If a student has that coin(s), have him or her place it face down on his or her board.
- The first person to have all six coins face down calls out "Coin Bingo."

### Variations

- The leader describes the coin by size, shape, and color rather than by value.
- Have students place their six coins face down and turn them face up if they match the leader's coin.
- Have students make a 3 squares x 3 squares Coin Bingo board for nine coins.
- Have students make a 4 squares x 4 squares Coin Bingo board for 16 coins.

# Coin Bingo

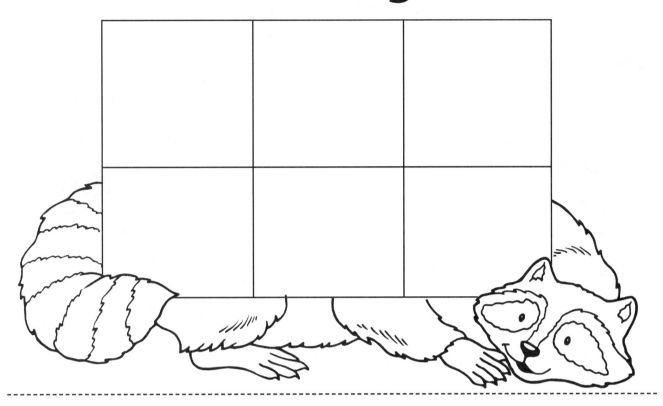

# Coin Bingo

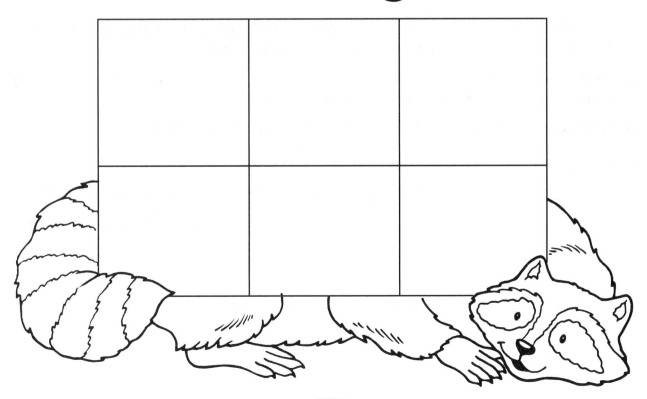

Exploring Coins

# Spin a Coin

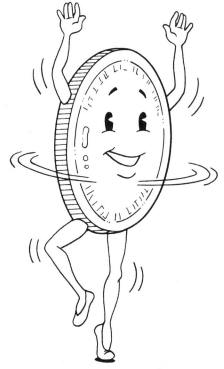

## Skills

- Recognize, name, and match 1¢, 5¢, 10¢, 25¢, 50¢, and $1 coins
- Describe and sort coins by design, color, shape, size, and value

## Grouping

- small groups

## Materials

- plenty of coins or paper coins (page 12)
- spinners (page 23)
- dice

## Directions

- Discuss the coins shown on the coin picture spinner.
- When it is each student's turn, have him or her spin the spinner and find that coin in his or her pile of coins.
- At the end of a time limit (e.g., five minutes), count to see who has the most silver coins. Ask, "Who has the most large coins? Who has the most five-cent pieces? Who has the most coins altogether?"

## Variations

- When it is each student's turn, have him or her throw a die. The number will indicate how many coins to collect. Have students use the spinner to find the type of coin they will collect. At the end of the time limit, have each student count how many coins of each type are in his or her pile. (e.g., Students can use the Sort Me cards on page 19 to sort their coins. Have them record their numbers on a piece of paper.)
- Have students use the spinner showing the numerical values in place of the coin pictures.

**Spin a Coin**  Exploring Coins

## Coin Picture Spinner

Color and cut out the spinner. Put a paper clip over the tip of a pencil. Place the tip of the pencil on the center of the spinner and spin the paper clip.

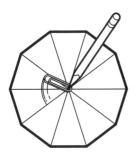
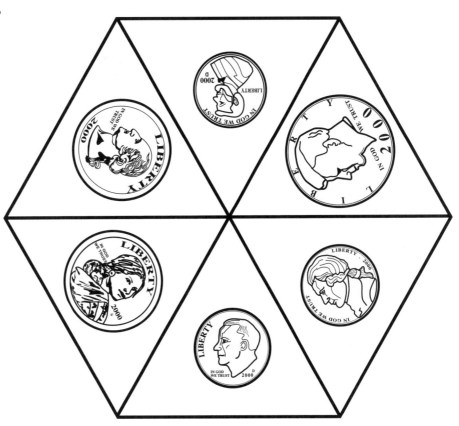

## Numerical Value Spinner

Color and cut out the spinner. Put a paper clip over the tip of a pencil. Place the tip of the pencil on the center of the spinner and spin the paper clip.

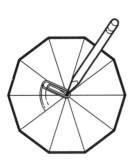
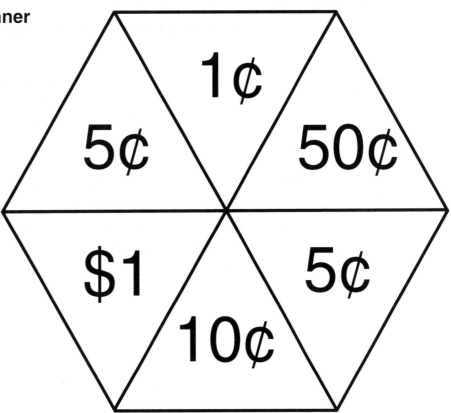

Exploring Coins

# Hide a Coin

## Skills

- Recognize, name, and match 1¢, 5¢, 10¢, 25¢, 50¢, and $1 coins
- Describe and sort coins by design, color, shape, size, and value

## Grouping

- pairs
- small groups

## Materials

- six coins (1¢, 5¢, 10¢, 25¢, 50¢, and $1)
- a paper bag

## Directions

- Have a student secretly take a coin and place his or her finger over the numerical value of this coin. Have him or her show it to a partner. Ask the students, "Can he or she guess the value just by looking at the shape/color/design that is showing?"
- Have a student secretly take a coin and place it in his or her partner's hand. Ask the partner to close his or her eyes. Ask students, "Can he or she guess the value of the coin just by feeling it?"

## Variations

- The leader secretly takes a random coin from a bag. The rest of the group or class try to guess the coin by asking questions. The leader can only answer "Yes" or "No" to each question. Ask students, "Can you guess the value in fewer than four questions?"
- Have a student place the six coins at random in a line. Ask the rest of the group to look at the coins and then close their eyes while the student hides one coin. Ask students, "Can you quickly guess the missing coin when you open your eyes?"
- Have a student place the six coins at random in a line. Have him or her hide two or more coins. Continue with guessing which ones are missing.
- Have each student make a coin pattern with a variety of coins. Have him or her ask someone to continue the pattern. Have the student hide one or more coins and ask a third student to guess the missing coin(s).

**Exploring Coins**

# Coin Concentration

## Skills

- Recognize, name, and match 1¢, 5¢, 10¢, 25¢, 50¢, and $1 coins
- Describe and sort coins by design, color, shape, size, and value

## Grouping

- pairs
- small groups

## Materials

- Coin Concentration cards (page 26 laminated and cut out)

## Directions

- Have students shuffle the cards and place the cards face down in rows and columns.
- Have each student turn over three cards at a time. If the cards match, the student keeps them. If they do not match, the student turns them face down again for the next player.
- Tell students the goal is to be the player with the most cards at the end of the game.

## Variations

- For a more challenging game, use two copies of the Coin Concentration cards (36 cards total).
- Have students play Coin Snap. Use two copies of the Coin Concentration cards (36 cards total) for each pair of students or small group. Have students shuffle the cards and deal out one card to each player until the deck is finished. Each student takes a turn to reveal his or her top card. The student calls out "Snap!" if his or her card matches the previous card revealed. The student keeps both cards if they match.
- Have students play Fast Match. Students will need a pile of mixed coins in the center. Have students shuffle the Coin Concentration cards. A leader turns over the top card. The other players race to find the matching coin. The first one to match the coin becomes the leader.

Coin Concentration | Exploring Coins

| | | |
|---|---|---|
| one cent | 1¢ |  |
| five cents | 5¢ |  |
| ten cents | 10¢ |  |
| twenty-five cents | 25¢ |  |
| fifty cents | 50¢ |  |
| one dollar | $1 |  |

**Exploring Coins**

# Which Is Worth More?

## Skill
- ❑ Order coins by value

## Grouping
- ❑ pairs
- ❑ small groups

## Materials
- ❑ a pile of coins/paper coins (page 12)
- ❑ Coin Concentration cards (page 26)

## Directions
- ❑ Discuss the fact that fifty cents is less in value than $1 or $2, even though the number sounds larger.
- ❑ Have each student take a handful of coins (about 10 coins).
- ❑ On a signal, have students race to sort these coins in order from the lowest value to the highest value.
- ❑ Have students shuffle the cards and place them in the center face down.
- ❑ Have each student take a card each and reveal it to his or her partner(s).
- ❑ Ask students to look at their coins and cards. "Who has the most money? Whose money is worth the least? Which student have the same amount?"

## Variations
- ❑ Have students sort all the coins/cards from the highest value to the lowest value.
- ❑ Have students sort mixed items from the class shop into order by value.

Checkup                                                                                           Exploring Coins

**Directions:** Draw a line to connect the front and back of each coin.

**Directions:** Draw a line to connect each design to its coin.

**Directions:** Circle the highest value coin.

**Directions:** Cross out the lowest value coin.

# Using Coins

**In this unit, your students will do the following:**

- ❏ Add coins of the same value
- ❏ Add two or more coins of mixed value
- ❏ Record coin totals using $ and ¢
- ❏ Identify higher and lower coin totals
- ❏ Identify coins needed for a given price
- ❏ Identify coins of equivalent value
- ❏ Calculate change with coins

(The skills in this section refer to the Skills Record Sheet on page 94.)

# Count with Me

## Skill
- ❏ Add coins of the same value

## Grouping
- ❏ whole class

## Materials
- ❏ a jar of 10¢ coins
- ❏ jars of 1¢, 5¢, 10¢, 25¢, 50¢, and $1 coins as needed

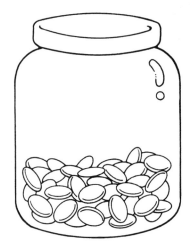

## Directions
- ❏ Have each student take a handful of 10¢ coins. Have the students guess how much money there is altogether. Have them count by 10s to check. (e.g., 10¢, 20¢, 30¢)
- ❏ Have students practice counting backwards by 10s, too. Have them put the coins they have counted back into the jar until there are none left in their hands. (e.g., 90¢, 80¢, 70¢)
- ❏ Have students repeat with a different handful of 10¢ coins.
- ❏ Ask students, "Did you collect more money this time?"

## Variations
- ❏ Have students count forward and backward using 1¢, 5¢, 10¢, 25¢, 50¢, and $1 coins.
- ❏ Discuss what happens when students reach one hundred cents. Ask, "Do you want to call this $1?"

**Using Coins**

# Piggy Banks

## Skill

- ❏ Add coins of the same value

## Grouping

- ❏ pairs

## Materials

- ❏ a pile of 10¢ coins (real, plastic, or coin cut-outs)
- ❏ a purse/wallet for each student (or use piggy banks on page 32, cut into two cards)

## Directions

- ❏ Have each student secretly place a number of 10¢ coins in his or her piggy banks. Have him or her show the bank to a partner. Have each student ask the partner to guess how much money he or she has altogether. Have students count by 10s to check.

- ❏ Have each student sit back-to-back with a partner. Have a piggy bank and a pile of 10¢ coins for each student. Call out two instructions for each to follow.
  (e.g., "Put in four coins." "Put in eight coins.")
  Now have each student count how much money he or she has in the piggy bank. Ask students, "Do you have the same amount as your partner?"

## Variations

- ❏ Have students place same number of coins in two or more piggy banks. Have them guess first and then find out how much money they have altogether.

- ❏ Have students use multiples of 1¢, 5¢, 25¢, 50¢, and $1 coins in place of 10¢ coins, as appropriate. Have students count by 1s, 5s, 10s, 25s, 50s, or 100s to find out how much money they have altogether.

**Piggy Banks**            **Using Coins**

**Piggy Banks**            **Using Coins**

# Match My Card

## Skills
- Add coins of the same value
- Identify higher and lower coin totals

## Grouping
- small groups

## Materials
- a pile of 10¢ coins (real, plastic, or coin cut-outs)
- page 34, cut into 10 cards
- tokens/counters

## Directions
- Have students shuffle the cards and place them face down in the center. Each student takes a card and collects that much money in 10¢ pieces.
- At the end of each round, discuss who has the largest amount of money and who has the smallest amount of money. Whoever has the largest amount takes a token.
- Ask, "Who has the most tokens?" at the end of the session.

## Variations
- Have students use 1¢, 5¢, 25¢, 50¢, and $1 coins in place of the 10¢ coins. Have students make new sets of cards to match.
- Have students use page 35. Discuss how much each toy costs. Have students color coins to match the price tags.
- Have students use page 36. The coins have been emptied from each piggy bank. Have students color the matching amount underneath each piggy bank.
- Have students use page 37. Tell them to imagine they sorted all the coins they saved into these rows. Have them write how much money they have in each row of coins.
- Have students use page 38. Have them look at the coins in each piggy bank and circle the highest value coin in each bank. Have them color the lowest value coin in each bank.

**Match My Card**     **Using Coins**

| | |
|---|---|
| 10 cents | 20 cents |
| 30 cents | 40 cents |
| 50 cents | 60 cents |
| 70 cents | 80 cents |
| 90 cents | 100 cents |

**Match My Card** — **Using Coins**

**Directions:** Color coins to match.

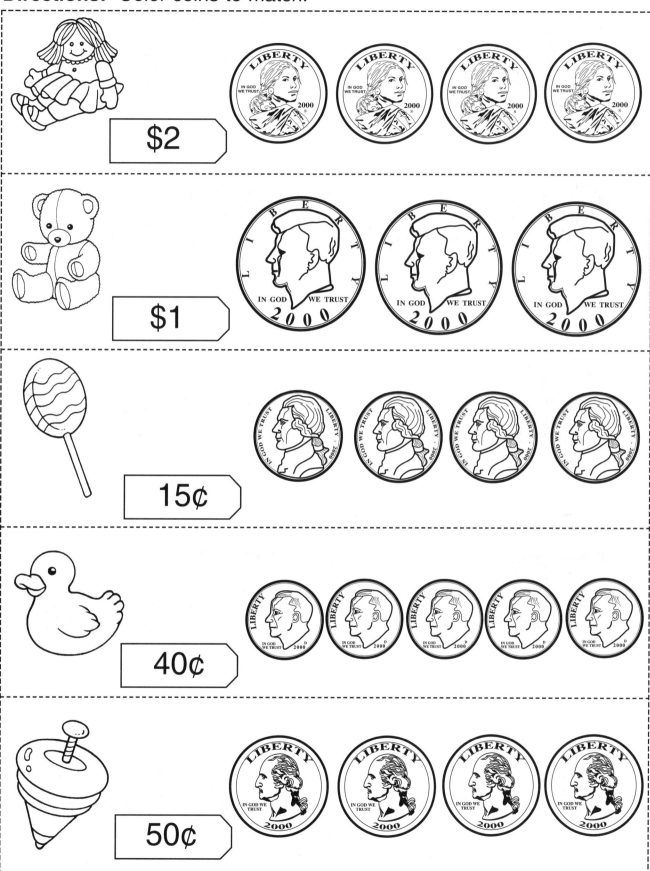

*Match My Card* *Using Coins*

**Directions:** How much money was in each money box?

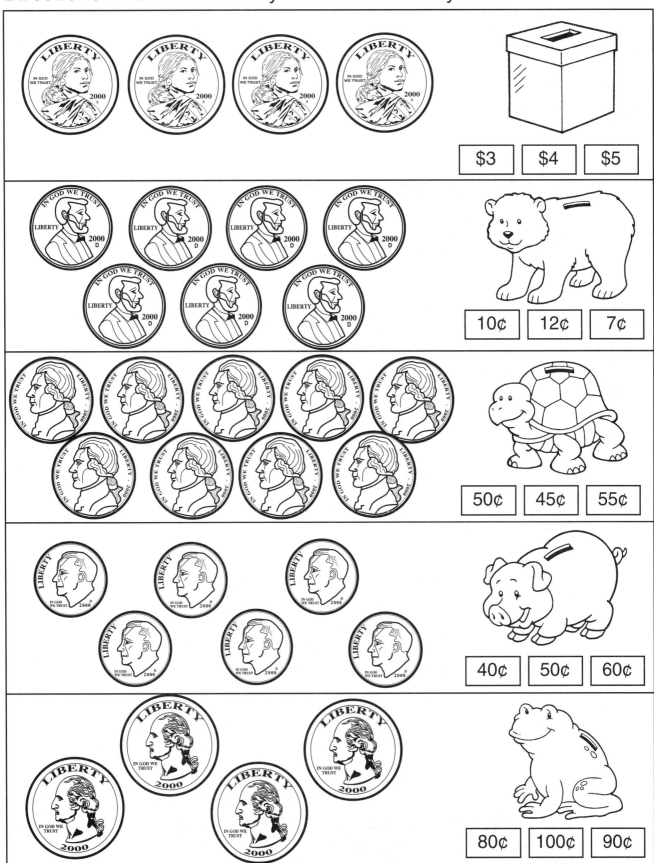

*Match My Card* *Using Coins*

**Directions:** Write how much money is in each row.

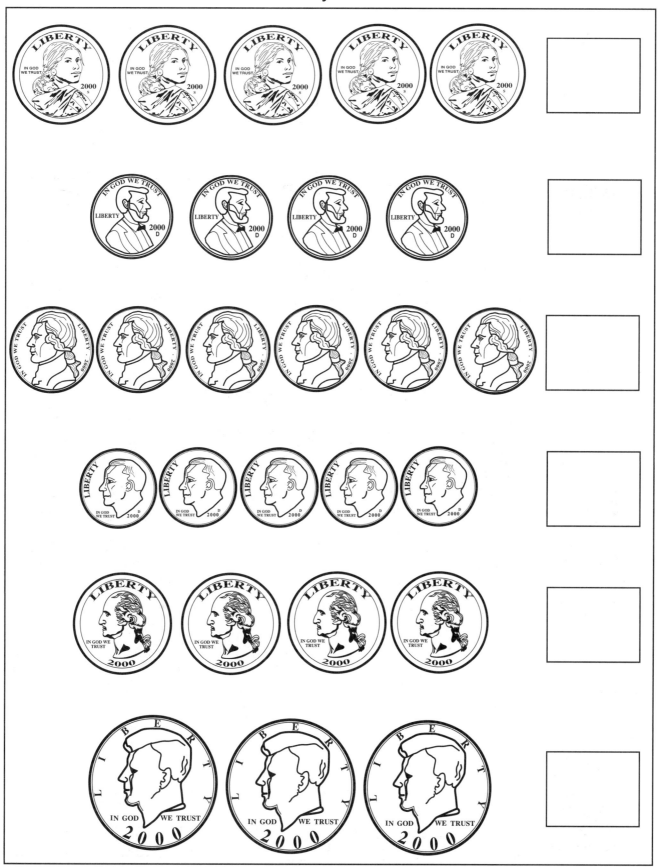

**Match My Card**          **Using Coins**

**Using Coins**

# Stack Them Up

## Skills

- Add coins of the same value
- Identify higher or lower coin totals

## Grouping

- small groups

## Materials

- a large pile of coins (real, plastic, or coin cut-outs)
- Stack Them Up cards (page 40, cut into eight cards)
- Coin Picture Spinner (page 23)
- Stack Them Up challenge cards (page 41)

## Directions

- Have students practice counting aloud by 2s, 5s, 10s, 20s, and 50s. If they get to 100, decide whether to call it "one dollar" or not.

    (e.g., 80¢, 90¢, 100¢, 110¢, 120¢)   or   (80¢, 90¢, $1, $1.10, $1.20)

- Have students take a handful of coins with the same value. (e.g., 10¢ pieces) Tell students to stack them up to make a coin cylinder. Have students guess how much money they have altogether. Have students check by counting (e.g., by 10s).

## Variations

- Have students use the Stack Them Up cards. Have them shuffle and place the cards face down in the center. Have them turn over the top card. Have them spin the spinner to see what type of coins are in this stack. (e.g., 25¢ coins) Have students guess how much money there is if the coins in the stack match this. Have them check by counting.

- Have students try the four Stack Them Up challenges, or invent your own.

    e.g.,   Which is worth more money,      I have a pile of 25¢ coins.
            a stack of four 25¢ coins        Altogether it is worth $1.25.
            or a stack of seven 10¢ coins?   How many coins do I have?

**Stack Them Up** **Using Coins**

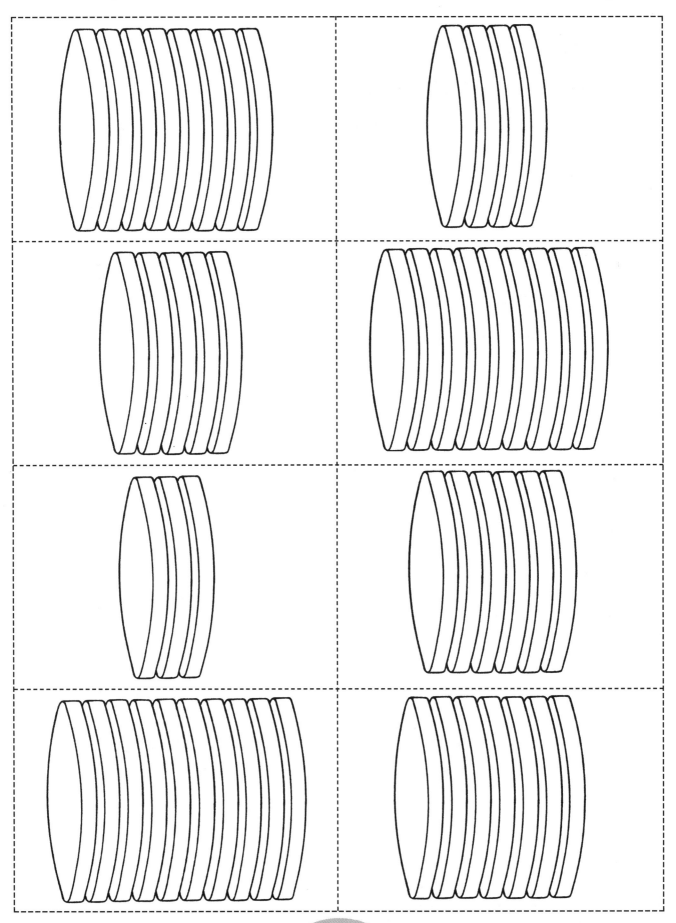

Which is worth more money,
a stack of five 25¢ coins
or a stack of nine 10¢ coins?
Guess first, then check.

I have a pile of 5¢ coins.
Altogether it is worth 45¢.
How many coins do I have?
Guess first, then check.

Which is worth more money,
two $1 coins or a stack of
twenty 10¢ coins?
Guess first, then check.

I have two 25¢ coins in my top pocket.
I have four 10¢ coins in my bottom pocket.
Which pocket is worth more money?
Guess first, then check.

**Using Coins**

# Mix Them Up

## Skills

- ❏ Add two or more coins of mixed value
- ❏ Record coin totals using $ and ¢
- ❏ Identify higher or lower coin totals

## Grouping

- ❏ pairs   ❏ small groups

## Materials

- ❏ a large pile of coins (real, plastic, or coin cut-outs)
- ❏ a purse/wallet for each player (or use banks on page 32)

## Directions

- ❏ Have each student take two different coins (e.g., a 25¢ and a 5¢ coin) and place them in his or her wallet.
- ❏ Have students exchange wallets with their partners. Have students find out how much money they have in their "new" wallets.
- ❏ Ask students, "Who has the larger amount of money?"
- ❏ Have students record their findings. (e.g., trace the two coins on paper and write the total amount, make columns for $ and ¢)

## Variations

- ❏ Discuss what happens when you mix $1 coins with a smaller value coin. Ask them, "What language do we use?"

  e.g.,  and    "One dollar and twenty-five cents"

  How can you record this?   (e.g., $1.25)

- ❏ Have students collect three coins of different values and find out how much money they have altogether. Have them record their totals.
- ❏ For a Super Challenge, have students collect four or more coins at random. Have them guess how much money they have and then check by counting.

*Using Coins*

# Supermarket Challenges

## Skills

- Add two or more coins of mixed value
- Record coin totals using $ and ¢
- Identify higher or lower coin totals
- Identify coins needed for a given price

## Grouping

- small groups

## Materials

- plenty of 1¢, 5¢, 10¢, 25¢, 50¢, and $1 coins
- a collection of supermarket grocery advertisements or use page 44
- pencils, paper, scissors, and glue

## Directions

- Have students make their own supermarket challenges using page 44 or a collection of supermarket grocery advertisements. Have them look for grocery items that have whole number price tags of $10 and under.

    e.g.,    one bunch of bananas  $2        juice  $3        soap powder  $8

- Have students cut out items for sale and glue them onto paper.
- Have students write a shopping challenge for another group to solve.

    e.g.,    What can you buy for $10?
                How much for two bunches of bananas and four bottles of juice?
                Record all the items you can buy for $15.
                How many different combinations can you discover?

- Have students use coins to work out the total price for their purchases.

## Variations

- Build up a class collection of supermarket challenges. Organize them by three levels of difficulty. (e.g., green—easy challenge, orange—challenge, and red—super challenge)
- Have students use page 45. Each basket shows grocery items bought in a supermarket. Have students write the total amount they will pay in the star.

Supermarket Challenges — Using Coins

**Supermarket Challenges**          **Using Coins**

**Directions:** What does it cost? Write the total amount for each basket.

*Using Coins*

# Find My Match

## Skills

❑ Add two or more coins of mixed value
❑ Identify higher or lower coin totals
❑ Identify coins needed for a given price

## Grouping

❑ small groups

## Materials

❑ plenty of 1¢, 5¢, 10¢, 25¢, 50¢, and $1 coins
❑ Find My Match coin cards (page 47)
❑ Find My Match word cards (page 48)

## Directions

❑ Have students shuffle the coin cards and place them face down in the center. Have them turn over the top card and say how much money altogether. Have students find the matching word card.

❑ Have students shuffle the word cards and place them face down in the center. Have them turn over the top card and collect the matching amount in coins. Ask them, "Is there more than one way to match that amount?" Have students find the matching coin card.

## Variations

❑ Have students check the amounts on the coin cards by matching with real coins.

❑ Have students turn over a coin or a word card each. Ask, "Who has the most money?"

❑ Have students use both sets of cards and lay them face down in rows and columns. Play Memory. Have students turn over two cards at a time and try to find two matches. Have students turn the cards face down again if there is no match.

❑ Have students write down the amount that matches either a coin or a word card (e.g., $2.70). Have them draw a picture of the coins to match.

❑ Have students use page 49. Ask, "Which combination of coins will you need to buy each toy?" Have them color the coins to match each price tag.

❑ Have students use page 50. Have them look at the coins in each group and look at the price tags. Then have them draw a line to join the price tag to the matching coins.

❑ Have students use page 51. Have them look at each row of coins and write how much money is in each row.

# Find My Match

## Using Coins

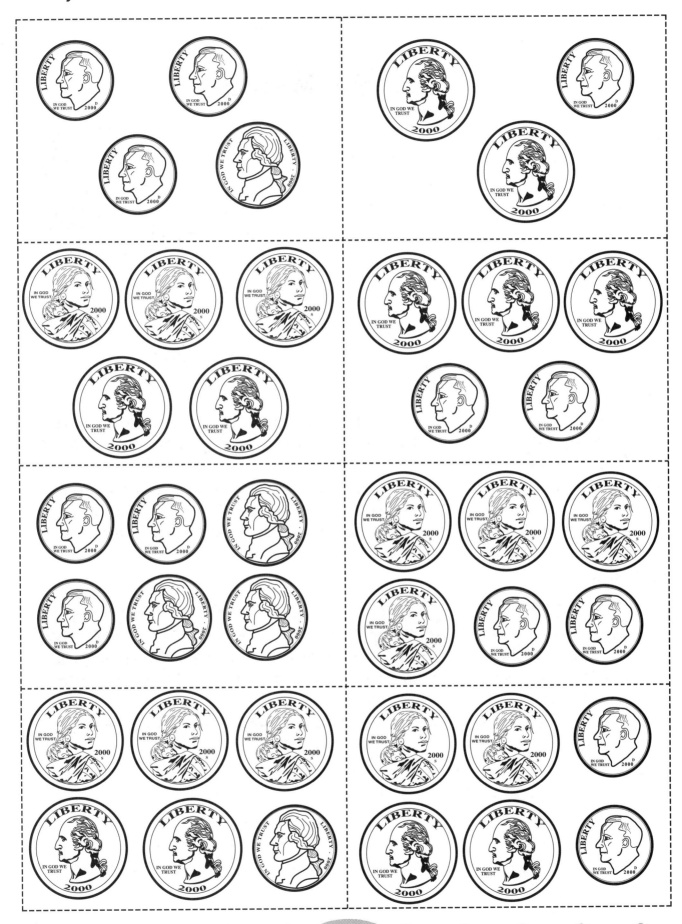

# Find My Match — Using Coins

**Find My Match** | **Using Coins**

**Directions:** Color the coins that match each price tag.

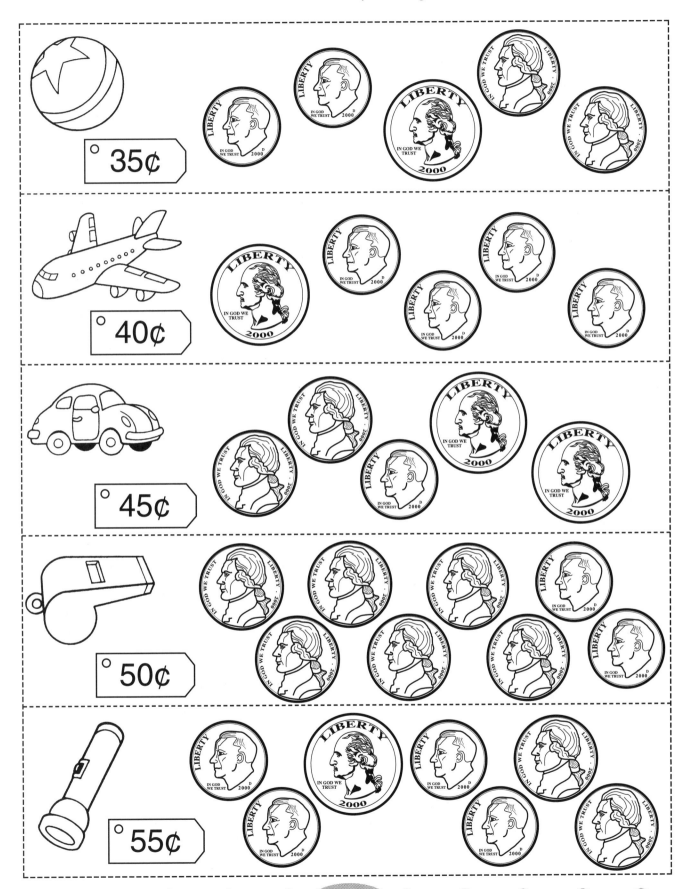

**Find My Match** **Using Coins**

**Directions:** Draw a line from the coins to the matching price tag.

**Find My Match** **Using Coins**

**Directions:** Write how much money is in each row.

# Class Shop

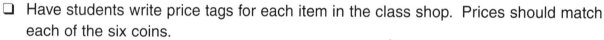

## Skills

- Add two or more coins of mixed value
- Record coin totals using $ and ¢
- Identify higher or lower coin totals
- Identify coins needed for a given price

## Grouping

- small groups

## Materials

- plenty of 1¢, 5¢, 10¢, 25¢, 50¢, and $1 coins
- a class shop (e.g., table, chair, shelves)
- mixed items to buy in the class shop
- price tags (page 53)

## Directions

- Have students write price tags for each item in the class shop. Prices should match each of the six coins.

e.g.,   1¢    5¢    10¢

 25¢    50¢    $1.00

- Have students practice buying several items at once. Ask them, "How much do you need to pay altogether? Which coins will you use?"
- Have students find a way to record their purchases.

## Variations

- Have students arrange the items in the shop from the lowest to the highest value.
- Have students give each item a price tag which uses two or more coins.

e.g.,   45¢    70¢    $1.30

- Ask students, "What could you buy with 75¢? Ten 10¢ coins? Five 50¢ coins?"

**Class Shop**        **Using Coins**

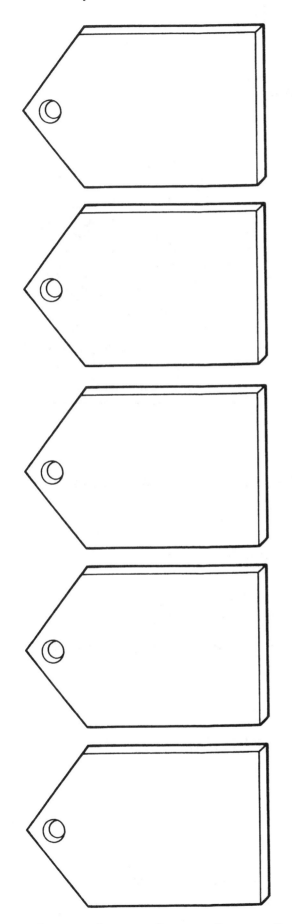
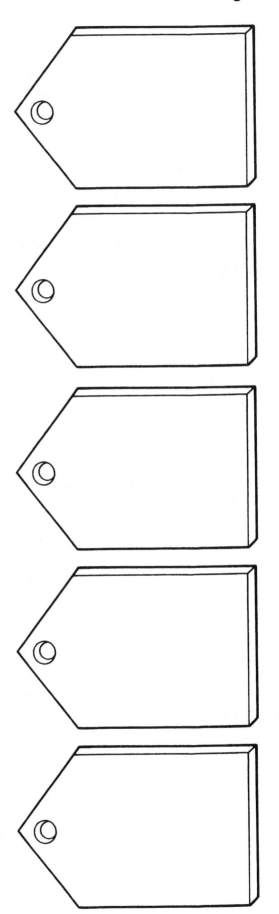

**Using Coins**

# Shopping with Up to 50¢

## Skills

- ❏ Add two or more coins of mixed value
- ❏ Record coin totals using $ and ¢
- ❏ Identify higher or lower coin totals
- ❏ Identify coins needed for a given price

## Grouping

- ❏ small groups

## Materials

- ❏ plenty of 1¢, 5¢, 10¢, 25¢, and 50¢ coins
- ❏ Shopping with Up to 50¢ cards (page 55)
- ❏ Shopping with Up to 50¢ challenge cards (page 56)

## Directions

- ❏ Have students imagine they are at the school fair. There are plenty of items for sale at good prices. Tell them to imagine each card represents one type of item. For example, there might be plenty of balls at 20¢ each, not just one.
- ❏ Have students shuffle the cards and place them face down in the center. Have them turn over two cards and look at the price for each item. Have them work out the total cost if they buy both items.
- ❏ Have students find the coin or coins needed to buy those items.
- ❏ Have students record what they bought and the amount they needed to pay.

## Variations

- ❏ Have students turn over three or more cards.
- ❏ Ask students, "What could you buy with 25¢? 35¢? 40¢? 50¢?"
- ❏ Have students try the Shopping with Up to 50¢ challenges. Have them make up more challenges like this for another group to try.

Shopping with Up to 50¢ — Using Coins

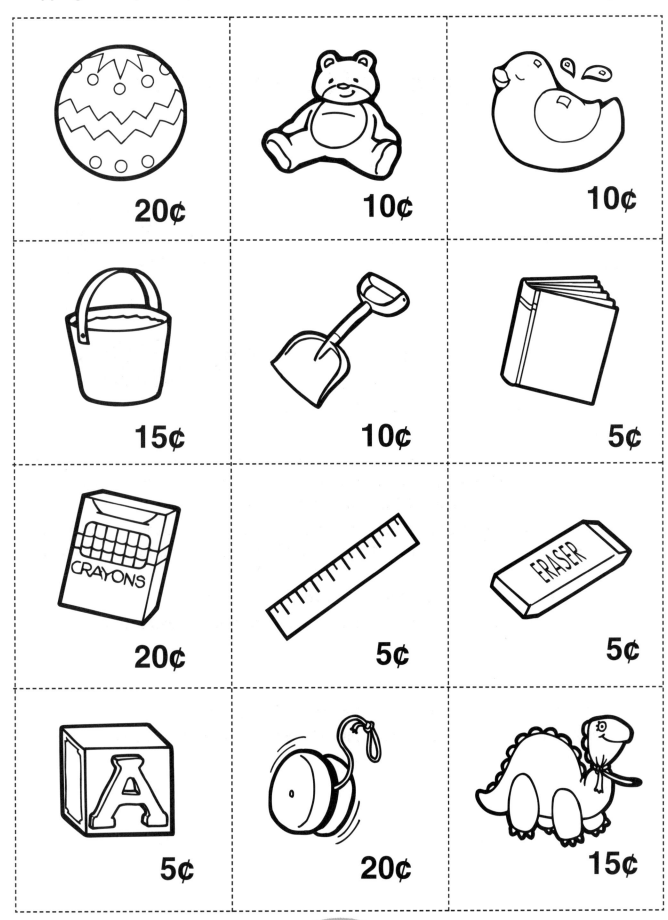

Find three items that together cost 50¢. How many different combinations can you discover?

Guess first and then check.

Could you buy three shovels with 50¢?

Guess first and then check.

Why? Why not?

I bought two items that cost me 25¢ altogether. What could they be? How many different combinations can you discover?

You have 20¢. Your friend has 20¢. Together how many books can you buy?

Guess first and then check.

**Using Coins**

# Shopping with Up to $1.00

## Skills

- Add two or more coins of mixed value
- Record coin totals using $ and ¢
- Identify higher or lower coin totals
- Identify coins needed for a given price

## Grouping

- small groups

## Materials

- plenty of 1¢, 5¢, 10¢, 25¢, and 50¢ coins
- Shopping with Up to $1.00 cards (page 58)
- Shopping with Up to $1.00 challenge cards (page 59)

## Directions

- Tell students to imagine they are at the school cafeteria. There are plenty of items for sale at good prices. Tell them to imagine each card represents one type of item. For example, there is a range of sandwiches at 45¢ each.
- Have students shuffle the cards and place them face down in the center. Have them turn over one card and say which coins they will use to buy this item.
- Have students turn over two cards. Ask, "How much do the items cost altogether?" Have them find the coin or coins needed to buy those items.
- Have students record what they bought and the amount they needed to pay.

## Variations

- Have students use a real school cafeteria price list. Have them find different combinations for a lunch order.
- Have students turn over three or more cards and find the total.
- Have students use the four challenge cards. Have them invent their own shopping challenges.

    e.g.,   What could you buy with $1? $1.50? $2?
            Find two items that together cost $1 or less. (e.g., a drink and a bag of marbles)
            How many different combinations can you discover?

Shopping with Up to $1.00 — Using Coins

**Shopping with Up to $1.00**     **Using Coins**

You want to buy three cans of soda. Will $1.00 be enough money?
Guess first and then check.

You have 75¢ to spend at the cafeteria. What can you buy?
Guess first and then check.

You have $1.00 to spend. How many slices of pizza can you buy to eat with your friends?
Guess first and then check.

You have 50¢. Your friend has 75¢. Together, how many frozen treats can you buy?
Guess first and then check.

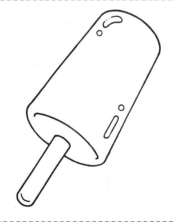

**Using Coins**

# Cash Registers

## Skills
- Add two or more coins of mixed value
- Record coin totals using $ and ¢
- Identify coins needed for a given price

## Grouping
- small groups

## Materials
- cash register for each player (page 61)
- four cash register price strips for each player (page 62)
- scissors
- glue
- items from the Class Shop or shopping cards (pages 44, 55, and 58)

## Directions
- Have students cut along the four dashed lines on their cash registers to make four slots.
- Have students cut out the four price strips.
- Have students glue the top end of the $10–$19 strip to the bottom of the $0–$9 strip. Have them thread this new $0–$19 strip through the left hand slots from behind their cash registers. Have students glue the ends together to form one long loop.
- Have students glue the top end of the 50–95 strip to the bottom of the 00–45 strip. Have them thread this new 00–95 strip through the right hand slots from behind their cash registers. Have students glue the ends together to form one long loop.
- They are now ready to record price totals on their cash registers by pulling the loops through until the price they want is revealed.

## Variations
- The leader selects two items from the Class Shop. Students race to show the total on their cash registers.
- Have students shuffle the shopping cards. Have them turn over two cards. Students race to show the total on their cash registers.
- Have students add more than two prices and show the total on their cash registers.

**Cash Registers** **Using Coins**

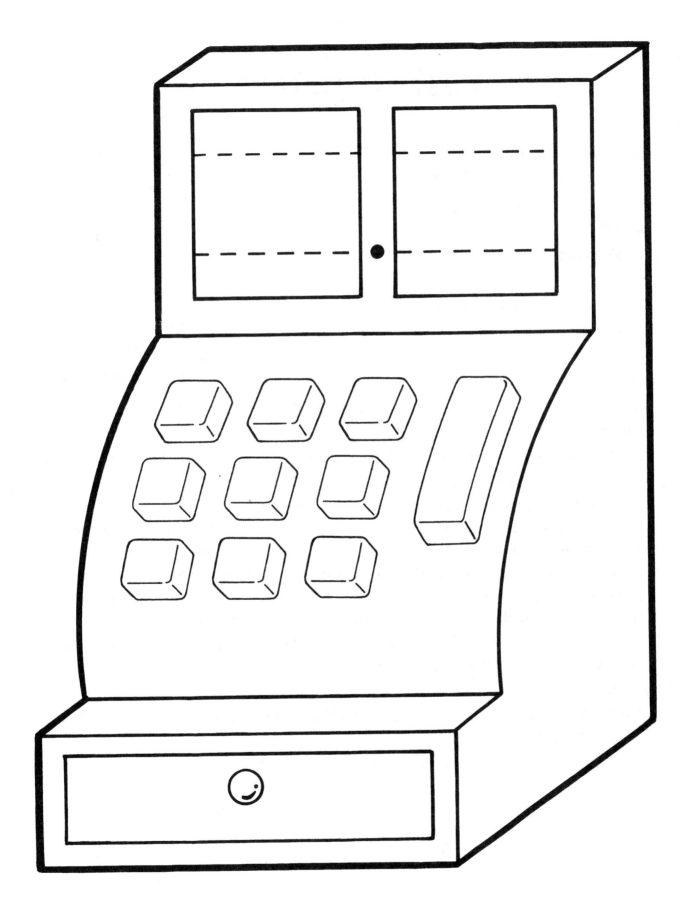

**Cash Registers**      **Using Coins**

| $0 | $10 | 00 | 50 |
| $1 | $11 | 05 | 55 |
| $2 | $12 | 10 | 60 |
| $3 | $13 | 15 | 65 |
| $4 | $14 | 20 | 70 |
| $5 | $15 | 25 | 75 |
| $6 | $16 | 30 | 80 |
| $7 | $17 | 35 | 85 |
| $8 | $18 | 40 | 90 |
| $9 | $19 | 45 | 95 |

Cash Registers     Using Coins

Using Coins

# Who Spent More?

**Skills**
- Add two or more coins of mixed value
- Record coin totals using $ and ¢
- Identify higher or lower coin totals

**Grouping**
- pairs

**Materials**
- calculators
- Who Spent More? worksheet (page 64)
- mixed items from the Class Shop (or use shopping cards on pages 44, 55, and 58)

**Directions**
- Have each student buy two or more items from the Class Shop.
- Have each student record his or her costs on his or her worksheet.

    e.g.,

- Have each student calculate his or her total cost by adding in his or her head, working it out on paper, or using a calculator.
- Have each student write his or her total cost on his or her worksheet.

    e.g.,

- Ask students, "Who spent the most money in your group? Who spent the least money? Who spent less than $2? More than $5?"

**Variations**
- Have students secretly take two or more shopping cards. Have them record the costs and find the total. Ask, "Who spent the most money?"
- Have students secretly select two or more items from a real shopping catalog. Have them calculate their total costs. Ask, "Who spent the most money?"

63

©Teacher Created Resources, Inc.   #3535 Math in Action

**Who Spent More?**        *Using Coins*

**Who Spent More?**        *Using Coins*

# Find the Same Amount

## Skill

- Identify coins of equivalent value

## Grouping

- individuals   ❏ pairs   ❏ small groups

## Materials

- plenty of mixed coins
- Find the Same Amount worksheets (pages 66 and 67)
- paper stapled to make a large class book about coin discoveries

## Directions

- Have each student take a coin at random. Ask, "Can you find two or more other coins that are worth the same amount?"

    e.g.,   10¢   "Can you find two coins worth the same amount?"

    20¢   "Can you find two coins worth the same amount?"
    "Can you find three coins? Four coins?"

    50¢   "Can you find two coins worth the same amount?"
    "Can you find three coins? Four coins? Five coins?"

- Have students record their discoveries. For example, students can draw, trace, cut out, or rub real coins on paper. Build up a class collection of all the different ways they can combine coins to make the same value.

## Variations

- Have students challenge other members of their class.

    e.g.,   "How many 10¢ coins are worth the same as $1?"
    "How many 25¢ coins are worth the same as $1?"

- Have students use the worksheets. Have them draw, trace, or cut out and glue coins to make up the total amounts in each section.

**Find the Same Amount**  **Using Coins**

**Directions:** Draw more coins to make $1.00.

**Directions:** Draw more coins to make 50¢.

**Find the Same Amount**        **Using Coins**

**Directions:** Draw enough coins so that the total value is $2.00.

**Directions:** Color those coins that will make $1.00.

**Directions:** Color those coins that will make $2.00.

67

*Using Coins*

# What's My Change?

### Skill
- Calculate change with coins

### Grouping
- individuals
- pairs
- small groups

### Materials
- plenty of mixed coins
- shopping cards (page 44)
- What's My Change? worksheet (page 69)

### Directions
- Tell students to imagine they have $10.00 to spend. Have students shuffle the shopping cards and turn over the top card. Ask, "How much change will you have if you buy that item?"

  (e.g., a bunch of bananas $2 ⟶ I'll have $8 left.)

- Have students find a way to record their actions.
- Ask students, "What if you bought two items?" Have students turn over two cards. Have them find out how much they need to spend. Ask, "Will $10.00 be enough money? How much change will you get from $10? From $20?"
- Tell students, "When you think you are ready and can figure out change from $10.00, use the worksheet. Look at each item for sale. If you had $10.00 to spend, how much change would you receive? Write this amount next to each item. Invent your own change challenge on the back of your worksheet."

*A bunch of bananas cost $2, so I have $8 in change.*

### Variations
- Have students reuse the worksheet later with a different amount to spend. (e.g., $15.00 or $20.00)
- Have students use the shopping cards from page 55. Repeat the activities, but this time they only have 50¢ to spend. Ask them questions such as, "What if you had 60¢? Would you have enough money to buy two items? What if you had $1.00?" Have students review the What's My Change? worksheet (page 70) with 50¢ (or more) to spend. Have them each write the change beside each purchase.
- Have students use the shopping cards from page 58. Repeat the activities, but this time tell them they have $1.00 to spend. Ask, "What if you had $2.00? $3.00? $4.00? $5.00? Review the What's My Change? worksheet (page 71) with $1.00 (or more) to spend.

**What's My Change?**            **Using Coins**

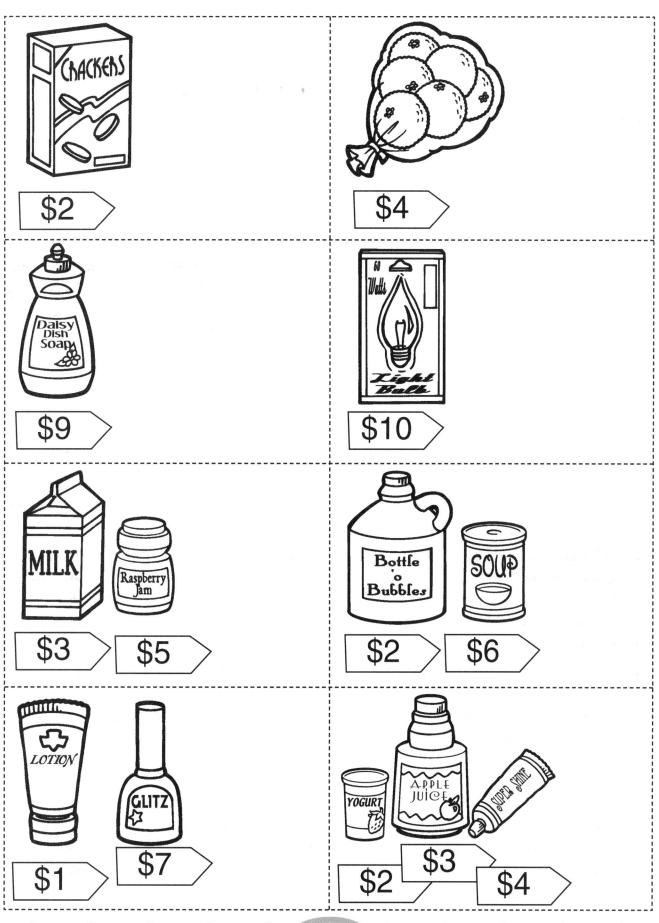

**What's My Change?** **Using Coins**

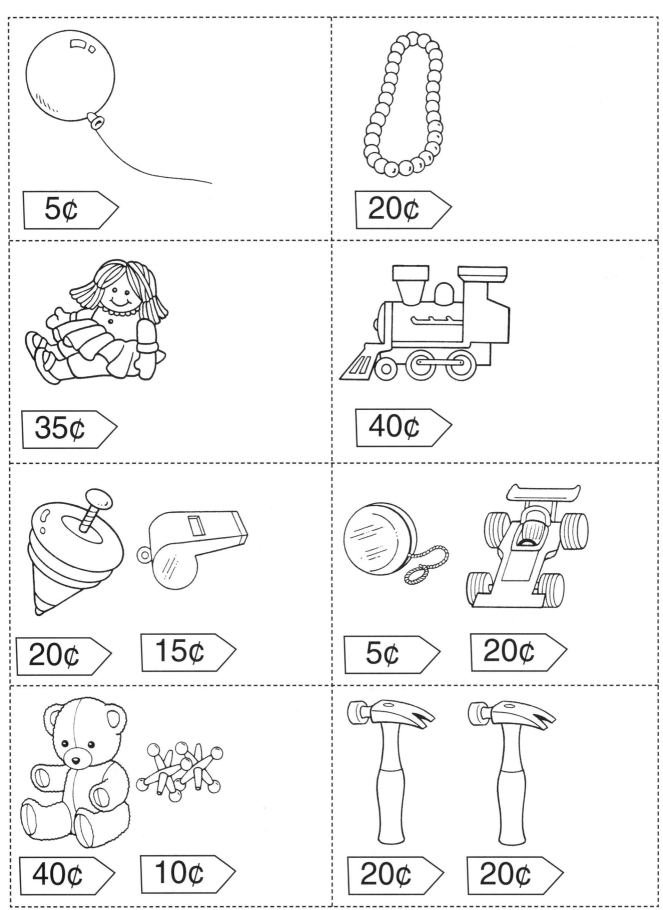

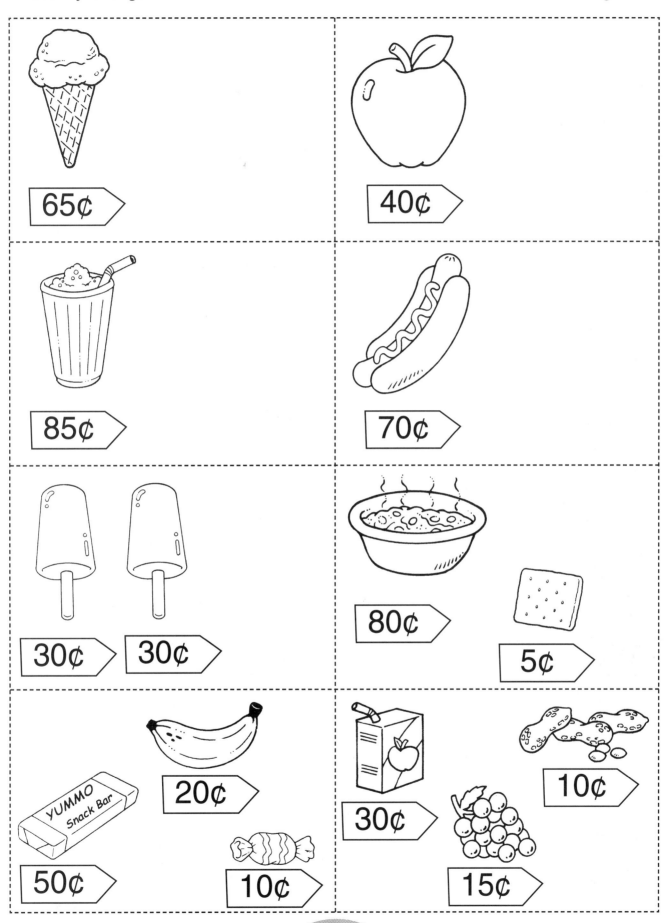

*Using Coins*

# Change Challenges

## Skills

- Add two or more coins of mixed value
- Calculate change with coins

## Grouping

❏ individuals   ❏ pairs   ❏ small groups

## Materials

- Change Challenge picture of a school fair (page 73)
- mixed coins
- pencils and paper for recording *(optional)*

## Directions

- Tell students to imagine they are at a school fair. Tell them to look at the picture with their partner(s). Discuss all the different things they can see. Ask students, "Which activity/item is your favorite? Why? Which one costs the most? Why?"
- Tell students to invent their own school fair change problem based on the picture. Tell them to imagine they have 50¢ or more to spend.
- In turn, have students challenge their partner(s) to solve their problems?

    e.g.,   Granny gave you 50¢. Can you buy two cakes and two toys?
    How much change will you have?
    You have $1.00. How much change will you get back if you buy
    six frozen treats to eat with your friends?

- Build up a class collection of written problems to go with this school fair picture. Challenge students in another class to solve the problems.

## Variations

- Tell students to use the toy shop picture (page 74) for Change Challenges with $2.00 or more to spend.
- Tell students to use the county fair picture (page 75) for Change Challenges with $5.00 or more to spend.
- Tell students to use the change cards (page 76). Have students shuffle them and turn over the top card. Have them invent a challenge based on two or more purchases from the school fair, toy shop, or county fair pictures.

Change Challenges — Using Coins

Change Challenges — Using Coins

Change Challenges — Using Coins

# Exploring and Using Bills

**In this unit, your students will do the following:**

- ❏ Recognize, name, and match $5, $10, $20, $50, and $100 bills
- ❏ Describe and sort notes by design, color, and value
- ❏ Order bills by value
- ❏ Add multiples of $5, $10, $20, $50, and $100 bills

(The skills in this section refer to the Skills Record Sheet on page 94.)

# What Is a Bill?

## Skills
- Recognize, name, and match $5, $10, $20, $50, and $100 bills
- Describe and sort bills by design, color, size, and value

## Grouping
- whole class

## Materials
- examples of $5, $10, $20, $50 and $100 bills

## Directions
- Discuss why people have bills for money. Ask students, "Why don't we just use coins?"
- Describe and discuss the designs they see on each side of a bill.

    e.g.,    $5

- Discuss the colors of each bill. Ask students, "What colors can you see?"
- Discuss the value of each bill.

    (e.g.,    What numbers can you see? Why does each bill have a different number?

## Variations
- Tell students to investigate the people who appear on each bill. Ask them, "Why do you think their portraits were selected? Why do we need to remember them?"

# Make Your Own Bills

### Skills
- Recognize, name, and match $5, $10, $20, $50, and $100 bills
- Describe and sort bills by design, color, size, and value

### Grouping
- individuals
- pairs
- small groups
- whole class

### Materials
- examples of $5, $10, $20, $50, and $100 bills
- United States bills (pages 80, 81, and 82)
- scissors, colored pencils, and glue

### Directions
- Tell students that they will investigate each bill further. Reveal just one corner of a bill. Ask students, "Can you recognize a bill just by looking at a tiny section? Is it possible to guess its value just by looking at the color?"

- Ask students, "What do you notice if you join the two short ends together to form a cylinder? *(They match exactly and join to form a continuous pattern.)* Why do you think they were designed this way? Does this happen with every bill?"

 e.g.,

- Ask students, "What do you notice if you join the two long ends together to form a cylinder? *(They also match exactly and join to form a continuous pattern.)*"

e.g.,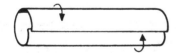

- Have students cut out the two United States bills pictures of the front and back of a $5 bill. Have them color them to match a real $5 bill's color. Have them glue the two sides together. Have them use these bills in their class shop. Have them repeat for each of the other bills.

### Variations
- Have students design their own notes. Tell them to think about the value, size, the color of the designs, and the pattern link-ups. Ask, "Who or what would you like to represent on your bill?
- Have students investigate the link between coins and bills.

*Make Your Own Bills*  *Exploring and Using Bills*

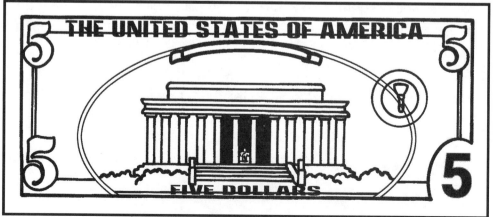

*Make Your Own Bills*          Exploring and Using Bills

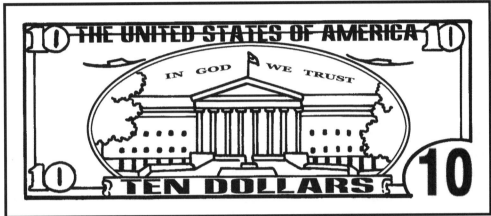
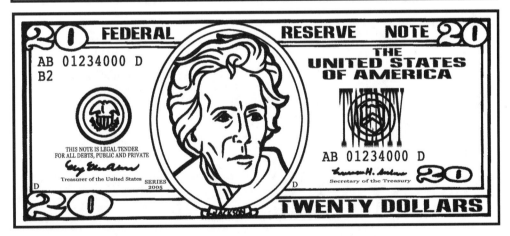

**Make Your Own Bills**    *Exploring and Using Bills*

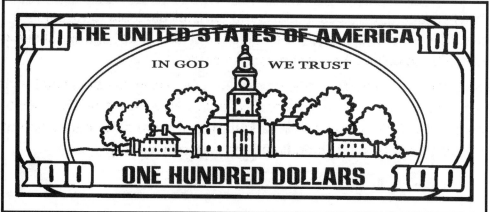

**Exploring and Using Bills**

# Which Is Worth More?

## Skills

- Recognize, name, and match $5, $10, $20, $50, and $100 bills
- Order bills by value

## Grouping

- individuals
- pairs
- small groups
- whole class

## Materials

- examples of $5, $10, $20, $50, and $100 bills
- home-made bills (pages 80, 81, and 82 colored, cut-out, and glued back to back)
- United States bills cards (pages 84 and 85)

## Directions

- Ask students, "Which of these two bills is worth more money? *(Hold up two different bills)* How do you know?" (e.g., $5 is worth five $1 coins, $10 is worth ten $1 coins)
- Have students shuffle the homemade bills. Have them select two at random. Ask, "Which one is worth more?"
- Have students shuffle the cards and select two at random. Ask, "Which one is worth more? Or are they the same value?" Have students try sorting three or more cards into order by value.
- Have students place the five different bills in order from the smallest to the largest values.

## Variations

- Have students play Note Snap in small groups. Have students shuffle the 20 cards and deal them all face down to each player. In turn, tell students to reveal their top card. If it matches the card before it, have the student call out "Snap" and win those cards. Tell students to try to collect the most matching cards.
- Have students play Note Memory in small groups. Have students shuffle the 20 cards and place them face down in rows and columns. Have students turn over two cards. If they match, the students keep them and have one more turn. If they do not match, have students turn them face down again for the next player to try.

**Which Is Worth More?**  **Exploring and Using Bills**

 $5

 $10

 $20

 $50

 $100

## Which Is Worth More?

**Exploring and Using Bills**

 five dollars

 ten dollars

 twenty dollars

 fifty dollars

 one hundred dollars

Exploring and Using Bills

# Make Your Own Catalog

### Skills
- Solve problems related to money
- Order bills by value
- Add multiples of $5, $10, $20, $50, and $100 bills

### Grouping
- individuals
- pairs
- small groups
- whole class

### Materials
- a large collection of supermarket, toy, or grocery shopping catalogs
- paper to make a book about United States bills
- scissors, glue, and pencils

### Directions
- Tell students to imagine they have $5.00 to spend. Ask them, "What do you know that costs about $5.00? Would you like to buy lots of low-priced items or one $5.00 item?"
- Have students look through the shopping catalogs. Have them find things they could buy for $5.00. Have them cut out their favorite ones and glue them onto a page. Have students draw a picture if they cannot find what they would like in a catalog. Build up a collection of things to buy for $5.00.
- Ask students, "How much would you need if you bought all your $5.00 collections?" Have students guess first and then check.
- Have students repeat, making their own catalog pages for $10.00, $20.00, $50.00, and $100.00 items. Ask, "Can you mentally add up the prices of several items at once?"

### Variations
- Have students ask friends to challenge them by calling out several items to buy, without them looking at their catalogs. Ask, "Can you remember the correct prices and add them all up to get the total? Can you mentally add more than four prices? More than five? More than 10?"
- Have students create their own shopping lists based on spending up to $100 from their personal shopping catalogs. Have them ask a friend to guess, then check how much money the items on one list will cost.

*Exploring and Using Bills*

# How Much Is That?

## Skills

- Solve problems related to money
- Add multiples of $5, $10, $20, $50, and $100 bills

## Grouping

- pairs
- small groups
- whole class

## Materials

- United States bills cards (pages 84 and 85)
- die (dots or numerals 1–6)
- paper and pencils (for recording)
- How Much Is That? cards (page 88, cut into 10 cards)

## Directions

- Have students practice counting by 5s, 10s, 20s, 50s, or 100s (as appropriate).
- Tell students to imagine they have lots of money and they like to count it aloud each day. Ask students, "Where would you keep it? What might you spend it on?"
- Have students shuffle the cards on page 88 and place them face down in the center. Have them turn over the top card. Tell them that this tells them what bill they have.

    e.g.,   $20

- Have students throw the die. Tell them that this number is how many bills they have of this type.

    e.g.,   5 x $20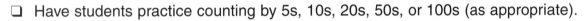

- Have students now work out how much money they have altogether.
- Have students find a way to record their actions.
- Ask students, "At the end of each round, who counted the most money?"

## Variations

- Have students shuffle the How Much Is That? cards and place them face down in the center. Have them turn over the top card and work out how much money that will cost altogether. Have them record it on their papers.
- Have students invent their own How Much Is That? cards for another team to use.

*How Much Is That?*                  *Exploring and Using Bills*

| | |
|---|---|
| Horse rides are $5 each. We had three rides. How much did we pay? | Movie tickets are $10 each. How much for six tickets? |
| Boxes of mangoes are $20 each. How much for three boxes? | A pair of new school shoes is $50. Mom buys four pairs. How much is that? |
| My dog has five puppies. Each one is worth $100. How much altogether? | Eight pies cost $40. How much for one pie? |
| Books were on special for $10 each. I spent $80. How many books did I buy? | We bought five kittens from the vet for $100. How much for one kitten? |
| I paid $100 for two computer games. How much did one game cost? | Grandpa paid $400 for four nights at a motel. How much for one night? |

88

# Pocket Money

## Skills

- Solve problems related to money
- Work cooperatively as a team
- Add multiples of $5, $10, $20, $50, and $100 bills

## Grouping

- pairs

## Materials

- Pocket Money problem (page 90, cut into six strips)

## Directions

- Discuss pocket money and how much students receive each week. Ask them, "On what do you spend it?"
- Ask students, "Do you save any of it?" Discuss how much money they might save in a week, a month, or a year. Ask, "How much might you have if you combined it with your friends' pocket money, too?"
- Have students look at the Pocket Money problem. The strips tell them a story about some friends. Have students discuss the problem in their own words. Ask, "How can you work out your answer?"
- Have students work together to find a solution. Ask, "What different strategies do you use?"
- Have students check their solutions against each statement. When students are convinced their solution is correct, see if they can discover another possibility.
- Have students invent their own problem about money for another team to solve. Tell them to try to make it have more than one solution.

## Variations

- Have students discuss their favorite ice cream. Ask, "Which flavor is the most popular? How much does it cost?" Tell them to try the ice-cream problem together (page 91). The price of each ice cream is different. Have students work together to find the solution. This time, there is only one way to solve it.
- Tell students to invent their own problem about money, like this one, for another team to solve.
- Tell students to try the puzzle cards (page 92, cut into six cards). Each card is a separate puzzle. Ask, "Can you find more than one solution to each problem?"

# Pocket Money — Exploring and Using Bills

Four friends saved their pocket money. They have six bills between them.

Lucy has less money than Duffy.

Duffy has more money than Eric.

Bev has much more money than Duffy.

Lucy and Bev each have two bills.

How much money could the four friends have altogether?

# Pocket Money — Exploring and Using Bills

We bought three different ice-cream cones.

We spent $4.50 altogether.

A strawberry ice-cream cone costs 50 cents more than a vanilla ice-cream cone.

A chocolate ice-cream cone costs double the price of a vanilla ice-cream cone.

A strawberry ice-cream cone costs 50 cents less than a chocolate ice-cream cone.

How much is a vanilla ice-cream cone?

I am a bill. I am worth more than $10 but less than $100.

We are two bills. We are worth less than $50 but add up to more than $20.

We are three bills. We are worth more than $70 but add up to less than $100.

We are four bills. We are worth more than $100. Three of us are the same.

We are five bills. Three of us are the same. Together we add to exactly $70.

We are five bills, worth less than $100, but more than $70. Only two of us are the same.

# Lots of Money

**Skills**

- ❑ Solve problems related to money
- ❑ Work cooperatively
- ❑ Add multiples of $5, $10, $20, $50, and $100 bills

**Grouping**

- ❑ pairs

**Materials**

- ❑ sixteen cards from United States bills cards (pages 84 and 85, excluding the four cards for $100)

**Directions**

- ❑ Have students practice adding the amounts shown on two cards. Have them each hold up two cards for their partners. Ask, "Can they add the two amounts in their heads?"
- ❑ Have students practice adding the amounts shown on three cards. Have them hold up three cards for their partners. Ask, "Can they add the three amounts in their heads?"
- ❑ Have students practice adding the amounts shown on four cards. Have them hold up four cards for their partners. Ask, "Can they add the four amounts in their heads?"
- ❑ For a Super Challenge, have students try this puzzle. Rearrange all 16 cards for $5, $10, $20, and $50 into four rows and four columns like this.

    e.g.,

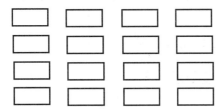

Make each row and each column add to exactly $85. Each row and each column must have a card representing $5, $10, $20, and $50 in it.

# Skills Record Sheet

NAME

## EXPLORING MONEY

| |
|---|
| Recognize the role of money in our daily lives |
| Recognize, name, and match 1¢, 5¢, 10¢, 25¢, 50¢, and $1 coins |
| Describe and sort coins by design, color, shape, size, and value |
| Order coins by value |
| Add coins of the same value |
| Add coins of mixed value |
| Record coin totals using $ and ¢ |
| Identify higher or lower coin totals |
| Identify coins needed for a given price |
| Identify coins of equivalent value |
| Calculate simple change with coins |
| Recognize, name, and match $5, $10, $20, $50, $100 bills |
| Describe and sort bills by design, color, size, and value |
| Order bills by value |
| Add multiples of $5, $10, $20, $50, and $100 bills |

# Sample Weekly Lesson Plan

**STRAND** Number  **SUBSTRAND** Money: Using Coins

**GRADE** 2  **TERM** 2  **WEEK** 8

## SKILLS
- Add two or more coins of mixed value
- Record coin totals using $ and ¢
- Identify higher or lower coin totals
- Identify coins needed for a given price
- Calculate change with coins

## LANGUAGE
- "This is the same value as . . . ."
- "This is worth more than . . . ."
- "These cost as much as . . . ."
- "How much change if I spend . . . .?"

## RESOURCES

1¢, 5¢, 10¢, 20¢, 50¢, and $1 coins (real/plastic/cut-out) pages 44, 55–56, 58–59, 61–62

Supermarket catalogs pages 45, 47–49, 64

Items for Class Shop pages 69–71

pages 73–76

| MONDAY | TUESDAY | WEDNESDAY | THURSDAY | FRIDAY |
|---|---|---|---|---|
| • Whole class: Review adding coins of same value (e.g. use goods from Class Shop) <br> • Mix Them Up (page 42) class discussion <br> • Activities: <br>  Group A: two coins (10¢, 25¢) <br>  Group B: two-three coins (any) <br>  Group C: four coins (any) <br> • Whole class discussion | • Review recording coin totals. (e.g., $1.20) <br> • Shopping Activities: <br>  Group A: 25¢, 50¢ coins (page 43, cards page 44) <br>  Group B: 5¢, 10¢, 25¢, 50¢ coins (page 54, cards pages 55 and 56) <br>  Group C: any coins (page 57, cards pages 58 and 59) <br> • Cash Registers (page 60) Whole class | • Shopping Activities: Group A: Free play at Class Shop (plus worksheet on page 45) <br> Group B: Find My Match on page 46 (plus worksheet on page 49 with teacher) <br> Group C: Shopping Challenge on page 57 plus Who Spent More? on page 63 (worksheet page 64) | • Whole class: How to estimate your change <br> • What's My Change? page 68 activities: <br>  Group A: from $10 (plus worksheet page 69) <br>  Group B: from 50¢ (plus worksheet page 70) <br>  Group C: from $1 (plus worksheet page 71) <br> • Whole class discussion | • Review week's activities <br> • Model Change Challenges (page 72) with whole class <br> • Change Challenge pair activities: <br>  Group A: page 73 <br>  Group B: page 74 <br>  Group C: page 75 <br> • Whole class discussion using cards on page 76 and "cash registers" |

# Weekly Lesson Plan

STRAND _____  SUBSTRAND _____

GRADE _____  TERM _____  WEEK _____

LANGUAGE _____

SKILLS

RESOURCES

| MONDAY | TUESDAY | WEDNESDAY | THURSDAY | FRIDAY |
|---|---|---|---|---|
|  |  |  |  |  |